William Stuart Watson

## Narrow Gauge Railroad System, a Complete Success

Its Adaptability to the Business of the Pacific Coast

William Stuart Watson

**Narrow Gauge Railroad System, a Complete Success**
*Its Adaptability to the Business of the Pacific Coast*

ISBN/EAN: 9783744681070

Printed in Europe, USA, Canada, Australia, Japan

Cover: Foto ©berggeist007 / pixelio.de

More available books at **www.hansebooks.com**

# NARROW GAUGE

# RAILROAD SYSTEM

### A COMPLETE SUCCESS.

---

## ITS ADAPTABILITY TO THE BUSINESS OF THE PACIFIC COAST.

---

COMPILED BY

## WM. STUART WATSON,

'CONSULTING ENGINEER,

Formerly Chief Engineer Buffalo and Pittsburg and Baltimore and
Pittsburg Railroads, the California Central, the California Northern, the Stockton and Copperopolis, the Yuba, San Francisco,
Central Pacific and other California Railroads. Member
of the American Institute of Civil Engineers, Etc.

---

SAN FRANCISCO, CAL.
APRIL, 1872.

# TO THE PUBLIC.

After having made a thorough examination of the NARROW GAUGE SYSTEM of Railroads now attracting so much of the attention of railroad men all over the world, and finding that the system has been an entire success wherever adopted, in furnishing a medium of transportation entirely adequate to the wants of the most densely settled countries, at a much reduced cost both for construction and maintenance, I propose to make the construction of such roads a speciality, and am prepared to make surveys and estimates of the cost thereof; to undertake their construction and superintendence; to furnish all the necessary rails and rolling stock at market rates, and to give all the required information on the subject necessary to a general introduction of the system.

WM. STUART WATSON,

P. O., 1749.

Office, 29 Merchants' Exchange, San Francisco.

# WHAT WILL A NARROW GAUGE RAILROAD ACCOMPLISH?

This question is daily becoming of more importance as the necessity for railroads becomes more generally admitted. It has been proven from the experience of railroad men in all countries that the railroads built on the standard gauge (of 4 ft. 8½ in. and upwards), have been vastly too expensive both in their construction and management; and although those countries through which they have been built have been enriched by their agency beyond all precedent. But experience has proved that shareholders have had to depend almost solely on the increased value of the lands through which the roads have been built; while the large amounts of subscriptions to the capital stocks that have been advanced by shareholders not similarly situated, have been almost an entire loss. If railroad men and capitalists who have given the subject any consideration with the light of experience before them, will look back to the early history of railroad construction, and trace their history, carefully from that time to this, they cannot fail to see that the system has been vastly more expensive than necessary, and will also become convinced that to continue the system must end in financial embarrassments, and result eventually in the ownership of nearly all the railroads already built going into the hands of foreign bond-holders.

Yet our undeveloped regions must be opened, and that they cannot be opened without the agency of railroads is as plain an axiom as the former is a fact. From such a statement the only conclusion that can be arrived at is that the present system must be changed for one much less expensive and adapted to the wants of the locality.

It ought to be engraved on the minds of every engineer and railroad promoter, that every inch added to the width of a gauge beyond what is absolutely necessary for the traffic to be handled, adds to the cost of construction, increases the proportion of dead weight, increases the cost of working, and in consequence reduces the useful effect of railroads.

The most important element in the organization of any enterprise is its probable cost; if the business required to be done will not warrant the expenditure, then no prudent capitalist will enter-

tain it. It seems to have been heretofore the
practice of railroad men generally to ascertain
how much money could be borrowed on the
securities of the road to be built, rather than how
small an amount the road could be built for, con-
sistent with the speed and safety of the transportation of its business. If the latter of these
propositions had been the ruling requirement in
the construction of our railroads instead of the
former, a totally new light would have been
thrown round the system and a totally different
result, financially, would have been arrived at.

The object herein proposed is to show in what
manner the cost of construction of railroads is
affected by the gauge on which a road is intended
to be built, and to what extent the maintenance of
a railroad may be and is affected by the gauge so
adopted; and to advocate from the experience of
the past such a system of railroads for the future
as will place this absolutely necessary element to
the development of any country, within the reach
of the most thinly populated districts.

And in order to make the subject as intelligible
as possible, and to place the facts as far as can be
ascertained before the public, I shall avail myself
of all the information withim my reach, which
will comprise reports from Great Britain, Germany,
Sweeden, Norway, Russia, Canada, the United
States and India, which I have taken the pains to

collect from the most reliable sources. In a paper of this nature it is scarcely possible to advance any new Idea, the ground has been so *often* and so *ably* covered, both by American and European engineers in their arguments during the " war of the Ganges," and since, that all the theory has been exhausted that the subject is capable of. It may be claimed however that the conclusions then arrived at were conclusions of *theory only*, which time alone could establish the truth of.

Nearly thirty years have elapsed since these theoretical arguments were had. During these thirty years the actual demonstrations of each theory has surely had time to be sufficiently developed.

Under these theories, about one hundred and twenty-two thousand miles of railroads have been built, and their actual yearly workings and earnings have become a matter of history in all countries. And we now find that the theory then advanced and limited to 4 ft. 8½ in. as a minimum gauge should have been extended to even 24 inches. Consequently the only argument that can or ought now to be advanced must be that of experience derived from the absolute working of the system which has by almost common consent been acted on for twenty years.

I therefore propose to avail myself of all the information that has been within my reach in bringing forward the argument that a *thirty-six inch gauge is all that is necessary* for any line of railroad on the Pacific Coast, or even in any country.

It will be necessary to go back to the commencement of the railroad era and review the arguments then held in the interests of the new proposed gauges, the wide (7 ft.) and the narrow, (4 ft. 8½ in.) The controversy occupied the attention of the ablest European and American engineers for nearly two years, and ended by the advocates of each system building their roads according to their individual theories.

## THE ORIGIN OF THE 4-8½ GAUGE.

George Stevenson advocated the 4 ft. 9½ in. gauge, because the wagons of England had axles of that length, and when the different parts of the "Rocket" were put together, owing to some unexplained cause, it proved to be only 4 ft. 8½ ins. The whole railway world then followed the example, because that engine made 14 miles per hour and the other competing engines made not more than half that distance, hence the present system.

I. K. Brunel, the most eminent engineer then living, brought forward the seven-foot gauge and the low flanges of the Great Western in England, which have long since been abandoned. Other engineers of high repute advocated all the means between these two extremes, but no one of them ever imagined that the smallest of their proposed gauges was wider than the experience of thirty years of operation would demonstrate to be the true one.

It is interesting to examine the conclusions that each party arrived at, and the special advantages that each set forth. It was argued that the broad gauge (7 ft.) was the most advisable. *First*, from the large capacity of the wheel base, consequent steady motion of rolling stock and a high rate of speed. *Second*, increased facilities for using more powerful locomotives, and, *thirdly*, a low centre of gravity for all rolling stock.

The objections were :

*First*—Increased cost of construction.

*Second*—Increased cost and weight of rolling stock, and increased liability of axles breaking, on account of their greater length.

*Third*—Increased friction of bearings in passing curves.

As regards the first of these objections, it was argued that the cost of construction was only increased in the 7 ft. gauge about seven per cent. in earth-works and land over the 4 ft. 8½ in. gauge. This, it will be remembered, was *theoretical* discussion only, as none of the eminent men engaged in the controversy had any experience in the

actual construction of railroads, and it is a curious
coincidence that the same arguments that were
advanced against the then narrow gauge (4 ft·
8½in.) and in favor of the wide gauge are identi-
cal with those now held against a reduction of the
present standard to three feet.

As to the objection of increased cost of construction in earth-works being only seven per
cent: Experience proved, during the construction
of roads under these two theories (7 ft. and 4 ft.
8½ in. gauges) that the cost of earth-works and
land damages on the wide gauge railroads
as compared to the cost of those items on
the Northwestern (a 4 ft. 8½ in. gauge) to be as
146 to 100.

As to the various points of objections held
against the wide gauge, both as regards increased
cost of construction, increased weight, and an
enormous increase in the operating expenses and
wear and tear of all portions of the equipment:
The advantage proven to be so evident in favor of
the 4 ft. 8½ gauge that within ten years of actual
operations no more wide gauge railroads were ever
constructed, and many that had been constructed
were reduced to the standard gauge as soon as
possible.

As regards the second objection held, then,
against a broad gauge, that of increased cost and
weight of rolling stock : It has been fully demon-

strated in practice that the locomotives and rolling stock of the standard (4 ft. 8¼ in.) gauge have been ample for all purposes of transportation and have been an entire success—except as to the cost of construction and maintenance, which time only could demonstrate.

From that time until within a few years it has been the practice of railroad men and engineers to take it for granted that a perfect gauge had been established, and whenever the subject has been at all discussed or questioned it has generally been concluded that it would be advisable to let the system remain in its present condition rather than to risk the negotiation of the securities of any important project by making a change in the system heretofore adopted.

Occasionally railroad companies advocating wide gauges in various countries of Europe, and also in America—for instance, the railroad system of Russia and such roads as the New York and Erie were commenced, and having been projected on a certain theory of construction, could not retrace their steps and were compelled to prosecute their works to a successful conclusion; but such as could retrace their steps without loss mostly have done so, many years ago.

Many instances have occurred where works of great magnitude have been commenced on the wide gauge system, the projectors have had the foresight to retrace the steps already taken and adopt the narrow gauge system instead.

Russia has abandoned the wide gauge system, and although their roads, most of them at least, that have been built, will continue to be worked for a time, yet all new roads being constructed in that country are being constructed on the small gauge system. Sweden and Norway are also reducing the width of their roads to a standard of 3 ft. 7 in. gauge, or to such a standard as the business may require, the gauge being the first consideration in constructing their roads.

For the sake of argument we will admit that the present system has satisfied all the requirements that was claimed for it, and that it has been the means of developing cheap and quick communication, as compared with the stage and lumber wagon, to an extent never dreamed of by its first projectors, increasing traffic and business to an almost unprecedented extent. Yet we find the objection raised that the returns for the capital outlay to stockholders is entirely inadequate; so much so that it is next to impossible to procure subscriptions to the capital stock of any railroad enterprise, however much needed or however promising the project may appear. Any promoter of a railroad enterprise knows from experience that the great objection he meets with is, that the stock of the road will be worthless; not because the project itself will not pay, but because the system compels the corporation to borrow money on its bonds at large rates of interest and at a ruinous discount. These bonds are negotiated at a fraction of their par value, and must be protected from the earnings of the road or a foreclosure will follow, the necessary consequence of which is that the stock is worthless, and that the roads built cost much more than they could be built for if built for cash subscriptions.

## WHAT THE STANDARD GAUGE HAS COST.

It has seemed heretofore that there was no other way to build railroads through the unsettled portions of the United States, except through the medium of issuing bonds, due years hence, at high rates of interest, the effects of which we now see in our enormous bond list and consequent high rates of transportation. If at the commencement of the construction of our present system, some plan more economical had been adopted, the bonded debt of our 60,500 miles of railroad, most of which has now but ten years to run, amounting to $2,200,000,000, would not now have existed, or if it had existed, would have been to a limited extent only, and would have long since disappeared, and railroads would have been clear of debt and paying dividends to the stockholders at thirty per cent. or more. And while this state of things has existed, and must exist until the system is changed, it is admitted by all those who are acquainted with the Pacific Coast, that hundreds of miles of railroads should and must be built to open our undeveloped regions. Yet the present system cannot be applied from its exorbitant cost; there are few settled countries that can bear the expense of transportation of the present system at $44,600 per mile; much less those partially settled, although known to be almost inexhaustible in wealth; and while knowing that fact, they also are aware that to attempt the development by the present system will entail a debt of such magnitude as to make it almost impossible for any prudent business man to make the attempt to carry.

It is claimed that on the grounds of first cost alone, if on no other, that this system of railroads should be abandoned, for some other much less expensive in construction, the first cost of which could be provided for by the section of country to be benefitted, without anticipating the revenues of the road for at least one generation.

There are many other and more cogent reasons why the system has outlived its usefulness and ought to be changed, which will be brought forward in the argument of what the system of railroads for the future ought to be.

If from the foregoing investigations of the

absolute workings, financially and otherwise, of these companies from whose reports the conclusions deducted here have been drawn, are anything like correct, then it is evident that any railroad or system of railroads that can be adopted on gauges from two to three feet in width will be productive of immense saving in cost of construction and economy in operating, and railroads would then be within the reach of sparcely settled countries, and would be especially adapted to the Pacific - Coast.

## STOCKS.

It has been ascertained that of all the railroad stocks issued for the construction of American railroads, amounting to nearly $2,500,000,000, less than one-fifth has a market value of par, and three-fourths or more of the whole are at a market discount of from ten to ninety per cent. Of the 550 railroad companies reported in 1871, not more than 200 of them have paid dividends on their stock during that year.

On these grounds alone, if no other can be adduced, it seems that railroad men should attempt to inaugurate some other system, which would secure to stockholders that return to which their expenditures entitle them, and which the business of the country demands and is compelled to pay for. It is evident that such a system can be found, and that it is being daily established in the reduction of the width of the present gauge. Lines of railroads are being constructed all over the world, on gauges of from 1 ft. 11½ in. to 3 ft. and 7 in. in width, which are giving entire satisfaction to their projectors and the public.

*As to the third requirement, that the stability in movement at such rates of speed as the dispatch of public business and the safety and comfort of passengers may demand.*

One of the principal arguments held against the narrow (4 ft. 8½ in.) gauge of 1844, and in favor of the 7 ft. gauge, was that the then narrow gauge (of 4 ft. 8½ in.), now in almost universal use, was defective in " wheel base," the absence of which, on account of the supposed lateral oscillation, would prevent a high rate of speed and endanger the safety of the passengers. It will be again observed that the same argument is held now, as against a reduction to any smaller gauge than that of 4 ft. 8½ in.

The experience of the last twenty years of the working of 4 ft. 8½ gauges has satisfactorily set at rest the objection on this ground, both as to an insufficient " wheel base " and oscillation. Indeed, the theory advanced then could not fail to establlsh the fact as claimed, and as will be proved now, that the oscillation of any train operating on a 4 ft. 8½ in. gauge, must be *absolutely less* than that incident to a 7 ft. gauge, and by parity of reasoning must be less on a small gauge road in the proportion as the small gauge recedes from the 4 ft. 8½ in. gauge, at least to the limit proved of 23½ inches.

[See diagram under head "Theory of Oscillation."]

In less than two years after the "War of the Gauges" ceased, the result of actual working of roads built on the opposing gauges was established to be true, as claimed by the advocates of the then narrow gauge.

Thus another of the supposed advantages of a "wide" over a "narrow gauge" was shown to be a fallacy.

The fourth and fifth requirements already referred to can probably be best satisfied by the question:

"What gauge will nearest fulfill the requirements of the railroad system in the future?"

In order to determine this question with any degree of certainty, we are happily placed in a more favorable position than were the scientific men who argued the question in 1841 to 1844, by having experience in the operations of the 4 ft. 8½ in. gauge (the wide gauge of our times) as to be able with a reasonable degree of certainty to apply a remedy for the defects of the present system, and to show what system will be

## THE RAILROAD OF THE FUTURE.

It appears plain that the railroads of the future must possess some other attraction than those of the past or of the present, and that they must possess the following, or some such capabilities :

First—That of smaller cost of construction.

Secondly—Smaller cost in maintenance and working.

Thirdly—To have the necessary stability, and to run at such rates of speed as the public dispatch, safety and comfort of passengers may require.

Fourthly—Capacity to do the business of the country through which they run; and,

Fifthly—To be available for all purposes in times of war.

With regard to the first and second requirements, the experience of railroad operations for the last twenty years shows most conclusively that the present system is a financial failure, caused principally from its unnecessary cost of construction and management—the result of its false system of finance and the unnecessary width of its gauge—in the same manner and with the same result as followed the seven foot gauge and its almost immediate abandonment for the narrower gauge **of 4 ft. 8½ in.**

**The** third, fourth and fifth of these requirements can best **be** answered **by** the actual operations of the narrow gauge roads already **built** and being built. We will therefore give a **brief account of those roads** that **have** already demonstrated the success of the **proposed** system :

## ROADS BUILT.

The Denver and Rio Grande Railroad, in Colorado, three **feet** gauge **and 117** miles in length, **which has proved an entire success.**

**The** Orangeville and **Toronto** Bruce Railroad, **in Canada, of 66** miles of three feet gauge.

The **Tifin road, in Ohio.**

The **Sodus Bay** Railroad, **in** the State **of New** York, and **the** projected system of 2 ft. **9 in.** gauges **in British** India, consisting of **from two to three thousand miles. Each of** these will be described under **their** proper heads.

**The 3** ft. **gauge, as** established **in** Sweden on a road of 23 miles **in** length, between Ullinbord **and the** Maclar seaport at Koping, on the Maclar sea, **which will** be hereafter noticed ; also the 3 ft. 6 in. gauge system adopted in Queensland, Australia, Ceylon, East Indies, Norway, and in Belgium, amounting to about one thousand miles already constructed, **where** they are found to give ample accommodation to the business of populous countries, and built with the capital of the country. The 2 ft. 6 in. gauges that have been built in Northern Germany, aggregating 640 miles, have fully satisfied all the requirements of business of that populous country; and lastly, the extreme of the proposed system, that of the Portmadock and Festinoig Railroad, in North Wales, which has been operated on a gauge of 1 ft. 11½ in. for 19 years with complete success. This road being, as may be supposed, the extreme limit to which the proposed system is likely to go, runs from the coast at Portmadock to the town of Dinas, in North Wales. It was originally built as a tram road, and the motive power was horses. The business is that of carrying passengers, slates and other freight between the seaport and quarries. It has been in operation for 19 years, and is now transporting 400,000 tons of freight and 150,000 passengers per annum. While with an original capital stock of $180,000, it has developed the industry for which it was intended, increased its capital stock to $500,000 and paid dividends of 40 per cent. on its original cost per annum.

## RUSSIAN COMMISSION.

It is probable that the best evidence that can be presented of the absolute success of this enterprise is the report made by the Russian

Commissioners, who were instructed to proceed to examine the workings of this road and to report to their respective governments. The commission consisted of the **Duke** of Sutherland, **accompanied by** Count Zchemji, of Hungary,

Count Alexis **Bobrinskoy, President** Russian Imperial Commission.

Count Zamoiski, **Count Berg,** Col. Statkowski, Imperial Engineer **of Russia.**

Professor Saloff, **Institute of Imperial Engineers of Russia.**

**M.** Raehrberg, Chief Engineer **and manager of** the Nigne-Moscow Railroad.

M. Schuberski, Superintendent **of locomotives of the Wornesch Rostou Railroad.**

**M.** Kislinski, **Russian** Imperial **Engineer and** Inspector of Karchof-Crementchay Railroad.

M. **Von Desen, M.** Sementechymoff, M. **Sach-**niffski, Russian **Imperial** Engineers **Serat Rail-**road.

Lieutenant-General **Sir William** Baker, R. E. **K. C. B.**

**Mr. W. T.** Thornton, Gorresponding Secretary Public Works, R. R. **and** Telegraph Department, Indian Office.

Mr. Juland Danvers, Government Director of Indian Railroads.

REPRESENTING BOARD OF TRADE.

Captain Taylor.

REPRESENTING FRANCE.

M. Gentry, President La Venda Railroad, M. Duval, Engineer La Venda Railroad, M. Krimer, Director Vote and Tiflis Railroad.

REPRESENTING SWEDEN.

M. Shandberg, Chief Engineer Swedish Government.

REPRESENTING NORWAY.

M. Pihl, Chief Engineer Norwegian Government.

REPRESENTING SWITZERLAND.

Mr. Carl Burkhardt, Chief Engineer.

REPRESENTING NORTH GERMANY.

Mr. Malrang, Engineer; Mr. Livingston Thompson, Managing Director Festinoig Railroad; Mr. Fleming, of Bombay, India; Mr. George Allen, Chief Engineer, India; Mr. Power, Vice President Vote and Tiflis R. R; Mr. G. B. Cranley, Contractor, Mexico; Mr. James Samuel, Chief Engineer and Director R. R. Construction Association, London; Mr. Toline, do; Mr. A. P. Hobson,

Secretary do ; Mr. **E. S.** Dallas ; Mr. Cargill ; Mr.
**Preston, London & N. W. R. R** ; Mr. Patchett,
Superintendent London & **N. W. R. R** ; Mr. Elias,
General Manager Cambrian **R. R** ; Mr. **Brough-
ton,** do Eid Wahs ; Mr. Walker, Locomotive Superintendent Cambrian R. R ; Mr. Roberts, **Chief
Engineer Brecon &** Merthyr **R. R** ; Mr. Caulfield,
do Neath & **Brecon R. R** ; Mr. Henshaw, General Manager Brecon & Merthyr **R. R.**

## REPORT.

### FIRST SERIES OF EXPERIMENTS.

Train hauled by a Farlie Engine, "Little
Wonder," 19 tons 10 cwt.

Feb. 11, 1870, started from **Portmadoc with the
"Little Wonder"** and

| | TONS. | CWT. | QRS |
|---|---|---|---|
| 90 Slate Wagons, weight | 57 | 10 | 0 |
| 7 Passenger cars and Baggage car | 13 | 10 | 0 |
| 57 Passengers | 4 | 00 | 0 |
| Locomotive and Tender | 19 | 10 | 0 |
| Total | 94 | 00 | 0 |

**Engine** Double **bogie 8 3–16 in.** cylinders x 13
in. stroke.

**Wheels 2** ft. 4 in. in diameter.

Pressure of steam **150** lbs. to the square inch.

Steepest grades 1 in. 74, (71.33 per mile.)

Shortest curves, 1¾ chs., (115.5 per radius.)

On the sharpest curves and steepest grades, the engine in full gear, the average speed was 14¼ miles, exclusive of stopping and starting. Length of train, 854 feet.

It was observed that on curves of 115.5 ft (1¾ chs) radius, and at a maximum speed of 30 miles per hour, there was very little perceptible oscillation or movement of the engine or in the cars, and by no means so much as is usually felt, even on comparatively easy curves on ordinary railways, and *less at high speed than at low speed.*

The super elevation of the outer rail on the sharpest curves was 3 inches.

(Feb. 12. Experiments with "Little Wonder," 19½ tons.)

With "Welsh Pony," 10 tons.

With "Mountaineer, 8 tons.

To test the steadiness of running on the Traethmawr embankment, on the "Welsh Pony" and "Mountaineer."

On the "Little Wonder," when riding on the foot-plates, no perceptible vertical or lateral oscillation, but a smooth, floating movement; when riding on the bogie frames, slight lateral oscillation, but less than on the other engines.

The oscillation of the Fairlie engine being confined to the bogie, the influence of impact on the rails, from the flanges of the wheels, was far less than in the case of the " Pony " or the " Mountaineer," the weight of these engines being brought to bear, in the course of their oscillation, upon the rails.

In all the above cases the speed was confined to 10 or 12 miles per hour, in a straight line, on a grade of 1 in. in 1,200 ft., or 4-40 ft. per mile, and the line was laid with rails weighing only 30 pounds per yard, not fished at the joints.

The " Welsh Pony " engine, weighing 10 tons, with cylinders 8½ in. in diameter by 12 in. stroke, and wheels 24 in. in diameter, took 50 wagons loaded with slate from Portmadock to the engine house, and stopped on a grade of 1 in 85, unable to proceed further, with 150 lbs. to the square inch steam pressure.

|  | | TONS. | CWT. | QRS. |
|---|---|---|---|---|
| Weight of 50 loaded wagons | | 123 | 10 | 0 |
| "      "      Passengers | | 3 | 10 | 0 |
| "      "      Engine | | 10 | 0 | 0 |
| Total | | 137 | 0 | 0 |

The " Welsh Pony " then took 25

| wagons, weight | 59 | 7 | 2 |
|---|---|---|---|
| Weight of Passengers | 3 | 10 | 0 |
| "      Engine | 10 | 0 | 0 |
| Total | 72 | 17 | 2 |

And mounted a grade of 1 in. 85 (or 62½ ft. per mile) with 145 lbs. steam in starting, and 130 lbs. after running about one quarter of a mile; then stopped to return.

The same engine then tried 30 wagons; could not start on a grade of 1 in. in 85 *but the engine wheels did not move:* there was, therefore, no want of adhesion, the load having been reduced to

|  | TONS. | CWT. | QRS. |
|---|---|---|---|
| 26 Wagons | 62 | 6 | 0 |
| Passengers | 6 | 10 | 0 |
| Engine | 10 | 0 | 0 |
| Total | 73 | 16 | 0 |

With an average pressure of 150 lbs. steam the "Pony" took them up a grade of 1 in. in 85 for a quarter of a mile at five miles per hour.

The "Little Wonder" left Portmadoc the same afternoon with 72 loaded wagons weighing

|  | TONS. | CWT. | QRS. |
|---|---|---|---|
| Slates | 138 | 17 | 2 |
| Empty Wagons | 43 | 13 | 0 |
| Passengers | 4 | 0 | 0 |
|  | 186 | 10 | 0 |
| Engine | 19 | 10 | 0 |
| Total | 206 | 0 | 2 |

And started with 165 lbs. of steam and ran to the engine house, and up the grade of 1 in. in 85 and was purposely stopped with steam at 125 lbs, pressure and a low fire, through the misapprehension of the engine driver. She was then backed

to the locality from which the " Pony " had started
with 26 wagons, and the fire being made up and
steam raised to 170 lbs. pressure, she freely started,
occasionally slipping, attained a speed of 5 miles
per hour with the 72 wagons, and after running
about a quarter of a mile she was increasing speed
on a gradient of 1 in. in 100 when she was
purposely stopped, with steam pressure still at 170
lbs. to the square inch.

In the above experiments the shorter trains were
standing, when they were started, or attempted to
start, partly on a curve of 2½ chs. (165 ft.) and in
the last experiment with the " Little Wonder," the
train having been longer, it stood partly on a
curve of 4½ chs. (297 ft.) and partly on a *reverse*
*curve* of 8 chs. (528 ft.) radius. The length of
this train was 648 feet.

The weather was fine, with a strong cold wind
blowing against the trains, and the rails were in a
remarkably good condition for adhesion.

The slate wagons had no springs ; the diameter
of their wheels was 1 ft. 6 in., and that of the
journals was 2¼ inches.

(Signed)       SUTHERLAND.
COUNT ALEXIS BOBRISKOY.
W. E. BAKER.
W. J. THORNTON.
W. H. TYLER.
JULAND DANVERS.

## SECOND SERIES OF EXPERIMENTS.

The result of an experiment on the Festinoig railroad on the 16th of February, 1870, with the "Little Wonder."

Length of Engine, 27 feet.

Weight of Engine in steam, 19¼ tons.

Diameter of Cylinders, 8 3-16 inches.

Length of Stroke, 13 inches.

Two 4-Wheeled Bogies.

Diameter of Wheels, 2 feet 4 inches.

Wheels coupled in each Bogie.

Wheel base of each, 5 feet.

Total wheel base, 19 feet.

### DESCRIPTION OF LOAD.

|  | TONS. | CWT. | QRS. |
|---|---|---|---|
| 22 Coal Wagons | 64 | 18 | 0 |
| 21 Wagons of Slate | 49 | 3 | 1 |
| 2 Bogie Timber Trucks, carrying timber 42 feet long | 4 | 18 | 2 |
| 15 Passengers | 1 | 1 | 2 |
| 2 Empty Trucks between Timber Bogies | 1 | 4 | 1 |
| 1 Workman's Carriage | 0 | 12 | 0 |
| Engine | 19 | 10 | 0 |
| Total | 141 | 7 | 2 |

Length of Train, with Engine, 407 feet.

The whole distance to be run over, from Portmadock to Dinas, 13¼ miles, having a total rise from a level of 703 feet, with maximum gradients of 1 in. in 92 for 12½ miles, (57½ ft. per mile), the Traethmawyr embankment near Portmadock being practically level.

The maximum curves are 1¾ chs. (115.5 ft.) Average curves, 6, 7, and 8 chs., the whole of the line being composed of a succession of curves, with the exception of the before named embankment and three or four other short portions.

The train started from Portmadock at 5:41 P. M.; at Penrhyn station at 5:58, without stopping there; arrived at Hartford Llyn station at 6:18, where it stobped 8½ minutes. Started at 6:26½, arrived at Ddwallt station (watering place) at 6:40, stopping 15 minutes on account of water having frozen, the tank could not be filled in the usual time.

The train reached the long tunnel at 7:02 P. M. through which it ran in 2 min. 10 sec. (Length of tunnel 730 yards, 2,190 feet.) Ran up to Tanygresion station, at which it arrived at 7:09 P. M., making the entire journey in 1 hour 34 minutes, including stoppages ; or exclusive of stoppages, in 1 hour 10½ minutes.

Maximum speed, 15 miles per hour.

Average speed, 11¾ miles per hour.

The engine, during the journey from Portmadock to Hartford, never slipped.

On starting from Hartford Llyn station a slight,

slipping occurred, (the **train being on a** curve of 4 chs.—264 ft—radius, with an inclination of 1 in. **100)** the rails being wet **and** heavy.

Slight slipping **on starting** from the watering place at Ddwallt.

The **engine** slipped **three** times in passing through **the tunnel, the rails being** wet throughout. Considerable **slipping took place at the junc-**tion of branch lines, this **place** being always wet and greasy, **owing to the** slate trains waiting **for** the down passenger trains.

The pressure of steam ranged from 160 to 180 lbs., at which **latter** pressure the train started, the pressure **being at** one time **only** 145 lbs. for a quarter **of a** mile.

Average pressure, 175 lbs.

The **entire** journey was run throughout **by the** engine **in only** two-thirds gear.

There was a **hard wind during the whole of** the journey, such being very **strong in some** parts of the line against the train.

(Signed)      LIVINGSTON THOMPSON,
                     C. E. SPOONER,
                     COUNT ALEXIS BROBINSKOY,
                     J. RAEHBERG,
                     PROFESSOR SALOFF,
                     R. VON DESEN,
                     J. SEMENTECHIMOFF,
                         Commissioners.

Under the superintendance of the President and Engineer of the La Vendee Railroad, of France.

Started from Portmadock with a train of 140 empty slate wagons and 7 loaded coal wagons.

Gross weight of load 100 tons, 16 cwt. 2 qrs.

Length of train 1323 feet.

Proceeded to Dinas, the upper end of the Festinoig railway. The maximum speed was 16¼ miles per hour; the average, 12½ miles.

Average gradients 1 in. in 92 ft., 57.4 per mile.

Average curves 5 60-100 chs., 375 24-100 feet.

On the return journey the speed attained was 30 miles an hour over many portions of the road, the average speed being 25 miles per hour.

## DESCRIPTION OF THE CONSTRCTION AND EQUIPMENT OF THE FESTINOIG RAILROAD.

The entire width of right of way procured was but 8 feet. The country through which it is built is extremely rough, there being but ten per cent. of the line tangent, with curves of 132 ft. radius. The iron first used when steam was applied was 16 lbs. per yard, which, with a traffic of about 350,000 tons per annum, lasted on an average of 14 years. The iron now used is 30 and 32 lbs. per yard. Speed attained, 40 miles.

Locomotives weighing from 10 to 19 1-2 tons haul about 300 tons of average loads. The passenger cars 12, 14, 16 ft. in length, by 6 ft. 7 in. width, carrying 14 passengers, or 200 lbs. dead weight for passengers.

Freight cars weigh from 1,100 lbs. to 1 1-2 tons and carry from three to four tons of load.

## E. C. SPOONER ON NARROW GAUGE FOR INDIA.

In this connection it may not be out of place to quote from the report of Mr. Spooner, Engineer of the Festinoig road, before the Inventors' Institute at London in 1865, on the system of small gauge railroads for India. He says:

" I have come to the conclusion that a 2 ft. 9 in. gauge is the most advisable for India, and it will fully meet all the requirements. And from my experience in working the 1 ft. 11 1-2 in. gauge I deduce the following, to show the sufficiency of a 2 ft. 9 in. gauge :

First.—That the cost in first construction in earthworks, bridges, tunnels, etc., depends almost entirely on the gauge. In regard to the construction, there is another matter for consideration with the Indian lines, which is of great importance ; namely : Being able to lay down a double line of 2 ft. 9 in. on a single line formation of 5 ft. 6 in.

gauge when required, without altering bridges, tunnels or earthworks, (as on all the Government lines, bridges, tunnels, viaducts, etc., are made for a double line of way) or going into any extra expense, except laying down a permanent way.

Secondly—That the cost of maintenance of rolling stock and permanent way will be low, consequent on the small weight on each wheel, and less damage to rolling stock in " shunting " or on collisions occurring.

Thirdly—That a speed of *forty miles* an hour can be run with *ease and safety*, as a speed of *thirty-five* miles per hour has been attained on a Festinoig railway. The present working speed is 16½ miles per hour, which is about the standard speed proposed for the Indian railways of 5 ft. 6 in. gauge.

Fourthly—As to the capacity of a 2 ft. 9 in. gauge to carry the required traffic. The Festinoig railway proves that a very heavy traffic can be conducted. The capabilities therefore of a 2 ft. 9 in. gauge in this respect must be apparent. To gain adhesion for the secured tractive power, to transmit heavy trains at necessary speed, and rolling stock of the required capacity, the most feasible known plan should be applied, namely : Engines of the Bogie principle, with four-wheeled double bogie frames, or six-wheeled double bogie frames ; and if need be, with lines of very heavy gradients and traffic, quadruple Bogie engines, by which means the weight of engines is distributed evenly—and rolling stock dispersed over the line of way "flange" and "drag" friction is reduced to a minimum, saving in wear and tear of rails and permanent way, and great advantage gained in passing sharp curves, all of which means money saved in maintenance of way and rolling stock, besides of fuel consumed for a given load.

The rolling stock can, by this means, be made to carry proportionately more than on the ordinary gauge. For instance, the bogie horse box weighs seven tons, to carry six horses, with compartments for groom and fodder, having when loaded 1 ton 6 cwt. on each wheel; whereas an ordinary horse box on a 4 8½ in. gauge weighs six tons and carries three horses, with compartment for groom, and having when loaded 1 ton 17½ cwt. on each wheel. The Bogie cattle truck weighs 5 tons 15 cwt. and carries ten cattle, weighing when loaded 11 tons 15 cwt., or 1 ton 9 cwt. on a wheel, while an ordinary cattle truck, on a 4 ft. 8½ gauge, carrying ten beasts, weighs 5 tons 10 cwt., and when loaded, 11 tons 10 cwt., or 2 tons 17 cwt. on a wheel. Surely these trucks, with a low centre of gravity and steadiness of motion secured, must have a far greater stability and freedom from oscillation than those in use on the ordinary lines. Bogie trucks can be made for the 2 ft. 9 in. gauge of any dimensions and convenient size for carriages for any purpose, of passenger or freight traffic."

And again I will quote from the same authority :

"The Festinoig railway carriages and trucks run very steady at a speed of thirty-five miles an hour, and prove to have the required lateral stability, even when going over the old rails, without "fish joints." There is no doubt that the rolling stock on the ordinary lines does not give nearly the carrying capacity compatible with the gauge. This can not be increased, as the weight already brought on each wheel is too great, causing the rails in *three or four* years, under a moderrate traffic, to become crushed or laminated, and under a heavy traffic at or near stations not lasting as many months. Steel rails have a greater durability—but at much greater cost—and even these scarcely sustain the great weight and impact blows of the heavy engines and rolltng stock, at high rates of speed ; besides steel rails wear out the tires much sooner than iron. The maximum weight on each wheel should not exceed three (3) tons, and can be reduced to two tons or under.

From the experience obtained in working small gauge rolling stock, the proportions that should be observed in the construction of trucks for engines and cars are regulated by the gauge, as follows : Width of trucks should be *two and a quarter* times the gauge; the depth *one and a half times;* the length, *four and a half times,* and the "wheel base" *two and a half times* the gauge outside dimensions."

## EXTRACTS FROM THE REPORT OF MAJOR ADELSKOLD, STATE NGINEER OF SWEEDEN.

From this report we select such portions as are pertinent to the subject, and illustrating the workings of a small gauge which lays between the standard 4 ft. 8½ in. and the 2 ft. 9 in. advocated for India, showing that the objection of a transshipment from one system of gauge to another is not attended with *any inconvenience* or much cost.

The road we refer to in the first instance is 23 miles in length, and runs between the seaboard at Maclar seaport to Ullinbord, through a district in which are several large iron works and sawmills, connecting that district with the small town of Koping, on the Maclar sea and the Royal Sweedish Main Line.

The gauge is 3 ft. 7 in., embankments 13 ft. wide, rails 37 lbs per yard, with fish joints, ties 6½ ft. long, 5 in. by 8 in.

The grades from the interior to the seaport are 1 in. in 200 in the direction of the heaviest trafic, and 1 in. in 100 in the opposite direction; maximum curvature, 100 ft. radius. The rolling stock was all built in Sweeden, and consists of three locomotives of 13 tons, 9 in. cylinders, 16 in. stroke, 4 wheels coupled, 2 ft. 3 in. diameter; 50 freight cars and 6 passenger cars.

The freight cars are of 6 tons capacity, 16 ft. long, 6½ ft. wide. The passenger cars are arranged for first and second class passengers, and carry twenty-six passengers—proportion of weight per passenger of 313 lbs.

The average speed is 16 miles per hour, and 35 miles per hour have been made on several occasions, making scarcely any lateral oscillation. The travel is not yet fully developed, and by doubling the number of cars 100,000 passengers and 150,000 tons of freight can be carried without any difficulty.

There are a number of other narrow gauge roads that have met all the requirements made of them, and entirely realized the expectations of their owners in every respect. Two of these are 4 ft. gauge; one 26 and the other 56 miles long, and are branches of the Sweedish Main Line. Prior to the completion of the first, it was generally believed that the transfer of freight from the branch to the main line would involve considerable outlay and serious loss of time. This objection has proved to be very trifling; the cost of transshipment does not *exceed one cent per ton.*

It was a subject of much doubt whether these small engines would keep the road open in winter.

The experience of several severe winters has shown that no fears need be entertained on this head.

He further says that "In every case where small gauges have been built they have realized every expectation," and deems it a waste of capital to build broader gauges when a narrower and cheaper one will meet all the requirements of business. In a thinly settled country, as this, where the available capital will hardly meet the wants of the rapid industrial and agricultural developments, and where cheapness of transportation is of the greatest importance, the smaller gauge may, in many localities, be well adopted, if not positively a necessity, as they can be easily built, and at such cost that with but a small traffic they are able, not only to cover the cost of operating them, but pay a good interest on the capital invested.

EXTRACT FROM ADDRESS BY MR. HULSE, PRESIDENT MANCHESTER INSTITUTE CIVIL ENGINEERS.

After discussing the present system on its financial theory, which has been in this paper already examined, with conclusions that the system has been much too expensive both in construction and management, he assigns the following reasons for a change from the present to a smaller gauge : That the 3 ft. 6 in. gauge is much preferable to the existing 4 ft. 8½ system for local railroads :

First—Because there is no necessity to use rails over 40 lbs. per yard; tunnels and bridges need not be of a hight exceeding 10 1-2 feet, earthwork of all kinds may be reduced at least one-third, curves can be used as sharp as 132 feet radius, locomotives need not exceed 15 tons, carriages not to exceed 3 tons and the weight on a wheel in no case need be more than 2 tons. Whereas on the 4 ft. 8 1-2 system the rails are 80 lbs. per yard, tunnels and bridges 24 ft. wide, open cuttings in proportion, curves not less than 570 ft. radius, locomotives 30 tons, carriages 8 to 20 tons, with a minimum load on each wheel of not less than 6 tons. The short curves that can be used on a narrow gauge avoid to a great extent the destruction of property, and can be adapted to the contour of almost any mountain country, reducing the cutting and embankment from through cuts and high embankments to " side-hill work."

He estimates that the 3½ ft. system costs less than two-thirds as much as the 4 ft. 8½ in. system, and can be worked at not to exceed two-thirds the expense.

This system has been largely adopted in Australia, Ceylon, Norway and Belgium with complete success.

The system of passenger cars adopted there is mostly what is know as the " Omnibus style," with seats arranged on the sides and a passage

30 to 36 inches in width down the middle, doors opening inward at each end.

The dimensions of such cars are : Length, 20 ft., width 6½ ft. and 6½ ft. high inside, Carriages of this size accommodate 24 passengers, twelve on a side, and give 30 cubic feet of space to each.

He further says that in some districts it might be found desirable to adopt 2 ft. gauges, for their cheapness, as has been done in North Wales.

EXTRACTS FROM THE CIRCULAR OF M. RIRD & CO., ON THE SPEED OF SMALL LOCOMOTIVES :

The narrow gauge locomotive, with driving wheels of 36 in. diameter, and cylinders 16 in. stroke, at a speed of 36 miles per hour, develops the same speed of piston as a full gauge locomotive with 5 ft. driving wheels and cylinders 24 in. stroke at a speed of 40 miles per hour.

With driving wheels 40 in. diameter and 16 in stroke of piston, the narrow gauge locomotive develops the same total travel of piston in going one mile as does a full gauge locomotive with driving wheels 60 in. diameter and 24 in. stroke of piston.

It is evident, therefore, that equal speeds are attainable on the norrow gauge as on the full gauge. The angle of stability of the narrow gauge locomotive, with 3 ft. driving wheels, is somewhat

greater than that of the 4 ft. 8 1-2 in. gauge locomotive, with 5 ft. driving wheels.

[See tables of comparative weights and dimensions of narrow gauge cars on closing pages of
this work.]

PATTERNS OF CARS IN USE ON NARROW GAUGE ROADS.

For passenger trains 8 wheeled cars with bodies
26 ft. long and 6 ft wide are proposed as follows :

1st.  First class cars, with revolving back seats,
(double seats on the one side of car and single
seats on the other, the plan being reversed in the
two ends of the car) to carry 28 passengers.

2d.  First class passenger cars, with two rows
of turning chairs and aisle between, to accommodate 18 first class passengers.

3d.  First class passenger cars, with cushioned
seats, like horse railway cars ; capacity, 30 passengers.

4th.  Second class passenger cars, with longi-
tudinal seats in centre (passengers sitting back to
back ;) capacity, 26 passengers.

All these cars to have wheels 24 inches in diameter.

PROPORTION OF NON-PAYING TO PAYING WEIGHT
TWENTY-NINE TO ONE ON PASSENGER TRAINS ON
THE STANDARD GAUGE.

It is now known, and everywhere admitted

among railroad men, that the proportion of non-paying weight in passenger trains to paying weight is as much as **19 to 1** in our ordinary first class cars, and in the sleeping and dining cars as much as **29 to 1**. In freight trains about 7 **to 1**. This terrible disproportion is partly **due to the** system of management pursued, **but in** a **far** greater degree to the **gauge.**

The dead weight of **trains** carrying either passengers or **merchandise** is **in direct proportion to** the gauge on **which** they run ; **or,** in other words, the non-paying to the **paying** weight (as far as **this is** independent of management**)** is increased **exactly as the rails are** placed farther apart ; for the simple reason that **a** ton of materials disposed **in** the **rolling** stock on a narrow gauge is stronger, **as regards its** carrying capacity, than the same **weight when** spread **over** a wider base. **In proving** this **we need** only **cite the Festinoig railroad.** The cars used there for **carrying timber** weigh only 1,344 lbs., **and** they **frequently carry** loads of 3½ tons, at **a speed of 12 miles per** hour. In other words, these **cars carry a** load as much as *five and a half* times **their own** weight, **while** the ordinary cars **on the** standard gauge carry not more than **one to one.**

In order to demonstrate this **part** of the argument, it **may** not be out of place to quote from the returns **of the Board** of Trade for Great

Britain in the year 1867, by way of comparison :
" The work done by the locomotives of Great
Britain in 1867, the last year for which returns
have been made, was 3,924,624 passengers trains
hauled 19.08 miles each, with a total tonnage of
Paying weight, 27,472,368, or 4.89 per ct.

Non-paying weight, 533,748,864, or 95.11 per ct.
And of freight trains, 2,403,866, hauled 30.64
miles each. With a tonnage of
Paying weight, 146,535,826 tons, or 30.34 per ct.

Non-paying weight, 336,541,240 tons, or 69.66
per cent.

And of total trains, 6,328,490 hauled 23 47-100
miles, with a tonnage of
Paying weight, 174,108,194 tons, or 16.67 per ct.

Non-paying weight, 870,290,104 tons, or 83.33
per ct.

Horizontal miles run :

Passenger trains.........................10,708,101,106
Freight trains ..........................14,804,545,302

Total........................25,212,646,408

This work was done by 8,619 locomotives, showing work done by each per annum of 2,960,047
horizontal mile tons ; work done by each per day,
8,966 horizontal mile tons—equivalent to 382 tons
hauled 23.47 miles per day, and 17,346 miles run
by each engine per year.

Taking the 23 47-100 miles, the actual average

distance run by each train per day, as consisting of an ascending grade of 1 in. in 300 ft. for half the distance, or 11,735 miles, and a descending grade for the remaining half, and assuming 26 miles per hour to be the average speed of each train, the following results are obtained:

Miles run each day per train, 23.47.

Tons weight of each train, 165.03.

Feet run per minute at 26 miles per hour, 2,288.

Lift of train per minute, $2\frac{280}{300}$, $7\frac{626}{1000}$ ft.

Horse power due to lifting, $85\frac{41}{100}$.

" " due to friction at 9 lbs. per ton, $102\frac{96}{100}$.

Horse power in ascending incline, $188\frac{37}{100}$.

" " in descending incline, 102.96—85.41–17.53.

Horse power exercised in a run of 23.47 miles, 205.92.

Trains hauled per day by each engine, 2.22.

Each train hauled $23\frac{47}{100}$ miles, at 26 miles per hour, gives $54\frac{16}{100}$ minutes occupied in the average journey, or for each engine an average working time of two hours per day.

The average horse power these engines are capable of exercising is probably not less than 400 hosse power each, making a total horse power of the 8,619 engines in use of 3,447,600, of which only 2,240,940 horse power can be made available. These results of course cannot be taken as

demonstrating the work that may be obtained
from any individual engine, but only as a means of
comparing the work done in one year with that
done in another in the same country, and in comparing the work done by locomotives in one country or one road with that done on another; and
above all, in showing how much the weakness of
the present system is responsible for, in taking
from the gross horse power which might be nearly
all economised in hauling paying freight. Hauling
dead weight not only never pays for its transportation, but wears itself and the rails out in one-
third the time that the actual business of the
country, economically managed, would require.

And again, if the dead weight of any train be
diminished, the weight of the engine may be similarly affected. And here is one of the strong
points of the narrow gauge system. It is evident
to any one who examines the rails of any of our
lines of railroads that have been in operation even
a short time, that the weight of the locomotive
and rolling stock is much too great, from the
crushed ends and the laminated appearance of the
tread of the rails.

The engines of the present system carry a
weight of from 4 to 6 tons, and even more, on
one wheel, while the actual business of the country can be as speedily and as safely carried with
engines, the weight of which are only 2½ or 3

tons to each wheel. In one case the weight of the motive power *crushes* the rails out of existence in five years; in the other they wear by attraction only and last fifteen years.

Aside from the locomotives of the present system being far too heavy, even the weight and power necessary for the system cannot be economically used to much more than one-half their capacity. I propose to examine why it is that passenger traffic cannot be utilized only in the proportion of one to nineteen. Let us look at the

## PASSENGER CARS.

One of our ordinary passenger cars weighs from 33,000 to 60,000 lbs., and the average of all through trains does not average 27 passengers for a 60 passenger car weighing, say 4,000 lbs. The non-paying weight therefore exceeds the paying weight from *eight to fifteen times.*

On a 3 foot gauge a car to accommodate thirty passengers weighs from 8,000 to 15,600 lbs., the passengers, as before, weigh 5,250 lbs., or from *one and a half to three times* the paying load of each car. In other words, the locomotive on the 3 feet gauge will have to haul from *twelve to twenty tons* lers weight in each car in the train than that of the wide gauge road ; but we will, for the sake of argument, say only seven tons less, and to keep the carrying capacity of the smaller road to the

standard of the wide gauge. business it may be
necessary to run an additional car capacity. Suppose, then, that five passenger cars are necessary
to do the business on an already established road
of standard gauge, and seven on that proposed for
trains of 3 ft. gauge. The weight of each train
will then be as follows:

WEIGHT OF PASSENGER TRAIN ON 4 FT. 8½ IN GAUGE.

|  | TONS NET. |
|---|---|
| 5 Passenger cars, at 33,000 lbs............. | 82 |
| 1 Baggage and Express car................. | 26½ |
| Weight of passengers and baggage........ | 10 |
| "    "    Engine and Tender............ | 45 |
| Total weight of train................. | 164½ |

ON A 3 FEET GAUGE.

|  | TONS NET, |
|---|---|
| 7 Passenger cars, 18,000 lbs., 126,000 lbs. | 63 |
| 1 Baggage and 1 Express car, 19,000 lbs. | 9¼ |
| Weight of passengers and bag'ge, 22,000 | 11 |
| Engine and Tender............................. | 17 |
| Total..................................... | 101¼ |

Assuming two trains *each way* per day for 313
days, there will be hauled

|  | TONS. |
|---|---|
| On the 4 ft. 8 1-2 in. gauge, gross tonage...205,954 |
| On the 3 ft. gauge, gross tonnage.............126,452 |
|  | 79,502 |

Showing a non-paying weight of 79,502 tons hauled on the wide gauge road; and suppose the actual cost of transportation is 2.56 cts. per ton per mile (which is not far from the cost in this state) we have an amount of saving per year of $2,035.25 per mile, or on 100 miles $203.525 in favor of the small gauge road in passenger traffic alone, aside from the saving in the wear and tear of the rails and rolling stock, which is nearly three to one.

And now let us look at the expenses in transporting freight on the two systems:

### CAPACITY AND WEIGHT OF FREIGHT TRAINS.

An ordinary 8 wheeled freight car on the standard gauge may be taken at 20,000 lbs., and carries a weight of about 18,000 lbs.

If the cars were fully loaded in each trip the proportion of non-paying to paying load would be as 10 tons to 9 tons; but we know that scarcely ever a freight car leaves any depot (except on long through journeys, and sometimes not even then) with a full load; and in every train that leaves and arrives at a terminus it is found that some of the cars have run empty between stations. For these reasons it is found that the non-paying weight carried by them is in the proportion of 1½ tons to each 1 ton of paying weight, or a 10 ton freight car cannot average a paying load of over 8 tons through the year.

On the other hand, a freight car on a 3 feet gauge, with a carrying capacity of 10,000 lbs., or 5 tons, and weighing 7,000 lbs., or 3 1-2 tons, will carry an average load the year round of 4¾ tons and still have a contingent capacity of 500 lbs. per trip for local business.

In this case the non-paying weight of 3½ tons has a carrying capacity of over 4¾ tons, or as 1 non-productive to $1\frac{46}{100}$ paying, against 1½ non-paying to 1 paying load

A freight train on the standard gauge, transporting 200 tons of paying freight will require 25 cars, weighing over 250 tons, which is 1¼ tons dead weight to 1 ton live weight; but on the small gauge 200 tons of freight will will require but 17 cars more than the standard gauge, or 42 cars in all for the trip, and will weigh 147 tons, or 1 ton of dead weight to $1\frac{36}{100}$ *paying load.*

### RESULTS OF COMPARISON—FREIGHT.

We have then the weight of a train transport- 200 tons of freight

#### ON A STADARD GAUGÉ.

|  | TONS. |
|---|---|
| 25 Freight Cars at 10 tons each | 250 |
| Freight Load | 200 |
| Weight of Engine and Tender | 45 |
| Total Tons of Train | 495 |

ON A 3 FEET GAUGE.

|  | TONS. |
|---|---|
| 42 Cars at 7,000 lbs. each | 147 |
| Freight Load | 200 |
| Weight of Engine and Tender | 17½ |
| Total tons of Train | 364½ |

Or 130 1-2 tons *less dead weight* to move the train on the 3 feet gauge than on the standard, and during 313 working days of 2 trains per day, one each way, the saving in tons hauled to do a business of 400 tons per day, or 125,000 tons per year, will average 81,690 tons in favor of the 3 ft. gauge and at 2 1-2 cents per ton per mile will show a saving on these two daily trains of $204,250 per annum.

In the transportation of minerals of all kinds the same or nearly the same results have been arrived at; that is to say, $\frac{3}{8}$ ton *dead weight only* carried to each ton of paying weight as against $1\frac{36}{100}$ tons of dead weight to one ton of paying load.

## COMPARATIVE REVENUE.

It has been shown that in four daily passenger trains, with a given business, the 3 ft. would save annually

| In working expenses of Pass. Trains | $203,550 |
|---|---|
| In    "    "    2 daily Ft. Trs. | 204,250 |
| Total saving on an annual business | $407,800 |

Which is **equal to an** interest of 5 per cent. on **six** and three-quarters **millions dollars.**

The business assumed for a **4 ft. 8 1-2 in** gauge 100 miles long, with double **track, we will suppose** to **cost** $8,000,000 for construction and **equipment, and** assuming the gross revenue to be of $2,000,000, deduct **60 per cent. working** expenses, $1,200,000, leaving net earnings, $800,000; or 8 per cent. dividend, and **2** per cent. sinking fund.

**A** railroad of like character **of 3 ft.** gauge will **cost not to** exceed $3,000,000; **will, of course, re-**ceive the same gross revenue **of** $2,000,000; allowing the **same** rate for working expenses, $1,-200,000, less **the** saving **shown** above, 407,800, leaving **for** net earnings $1,207,800, or a net profit of over **40** per **cent.,** or $3,000,000; or in other **words,** that while the standard gauge roads cost to **operate** them 60 per cent. of the **gross earn-ings** and leave but 40 per cent. for **revenue, the 3 ft.** gauge **can be operated at a cost of 26 per cent. of the** gross **earnings and leave 74 per** cent. for net revenue.

## NEW YORK CENTRAL, NEW YORK & ERIE AND OTHER RAILROADS.

Again **let** us examine the operations of two of the leading lines of **the** United States: The New

York and Hudson River, the New York and Erie, and other large roads:

It will be observed by any one travelling on our leading railroads that the passenger cars reserved for local business on through trains are but partially filled during many hours of the day, and an average in our 60-passenger cars of 27 to 30 passengers is considered good work. These trains generally move with about six passenger cars, weighing, with the engine, loaded tender and baggage car, about 360,000 lbs.—180 passengers to 360,000 lbs., gross weight of train, or 2,000 lbs. to each passenger; that is to say, for every ton of passengers that pays we carry fourteen tons dead weight without pay.

In the year 1871 the passenger traffic of the New York Central and Hudson River Railroads amounted to 321,365,953 passengers carried one mile; but they carried along with them 477,-750,000 tons of dead weight; that is, a ton and a half of dead weight to each passenger.

NEW YORK AND ERIE, 459 MILES OF MAIN LINE.

The New York and Erie Railroad, a 6 ft gauge double track, did worse than the New York Central. The number of passengers carried one mile is given at 135,589,100, with 448,250,000 tons of dead weight, which is three and one-third tons of dead weight per passenger.

In the freight traffic the New York Central moved 469,087,777 tons one mile, and 405,500,000 tons dead weight, or 86-100 of a ton dead weight for every ton of paying freight.

And the Erie moved 809,862,718 tons of freight one mile with 1,177,250,000 tons of dead weight, or $1\frac{32}{100}$ tons dead weight per ton of paying freight.

ATLANTIC AND GREAT WESTERN, 387 MILES MAIN LINE.

The Atlantic and Great Western Railroad carried 54,139,269 pass. one mile, and 179,060,000 tons dead weight, or three tons and thirty one-hundredths dead weight per passenger; while there was transported 252,353,696 tons of freight one mile, and 472,686,800 tons dead weight, or nearly two tons dead weight to one of paying freight.

CHICAGO AND NORTHWESTERN, 834 MILES MAIN LINE.

The passengers carried on this road for the year 1871, amounted to 100,802,512, carried one mile, with 139,980,000 tons dead weight, which is 1 39-100 tons dead weight per passenger; while the freight moved on this road during the year amounted to 268,417,881 tons, one mile, with 314,460,000 tons dead weight, or 1 17-100 tons dead to 1 ton paying freight.

It is needless to adduce more examples of the waste of motive power of the present system; there is, however, one more out of the many that the tables of last year show that we will quote, and then leave the subject: the

## LAKE SHORE AND MICHIGAN SOUTHERN RAILROAD,

Running through the States of New York, Pennsylvania, Ohio, Indiana and Illinois, a distance of $1,282\frac{8}{10}$ miles of line, consisting of $539\frac{9}{10}$ miles main line, $473\frac{1}{10}$ branch and $269\frac{8}{10}$ second track.

On this road, traversing portions of five states, the passenger traffic is reported at 159,390,937 passengers carried one mile, and 569,833,666 tons of freight, with the proportions in passenger traffic of $\frac{9}{10}$ tons to each of dead weight, and of freights $1\frac{22}{100}$ dead to 1 of paying weight; or for one ton of paying passenger freight carried, the road carries nearly 14 tons of load.

## THEORY OF OSCILLATION.

It has been asserted in another part of this paper that the lateral oscillation on a narrow gauge is *absolutely less* than on a gauge of 4 ft. $8\frac{1}{2}$ inches or over. The best illustration of this subject is taken from the able report of Mr. Sears, Chief Engineer of the Pennsylvania & Sodus Bay Railroad, in a report recommending a 3 ft. gauge for that company in 1871. He says:

" These systems have ceased to be experimental only—on the contrary, so successful have they become, that even in the United States where railroads are built the cheapest, the public attention is being drawn to them, and prudent men are asking why every section of the country cannot have its own railroad, when they can be provided at such cheap rates."

He says, on the matter of stability at high rates of speed :

It is feared by some that a 36-inch gauge will not give as stable equilibrium to our cars as is necessary, and asks the attention of his company to the accompanying diagrams, designated Fig. 1 and Fig 2.

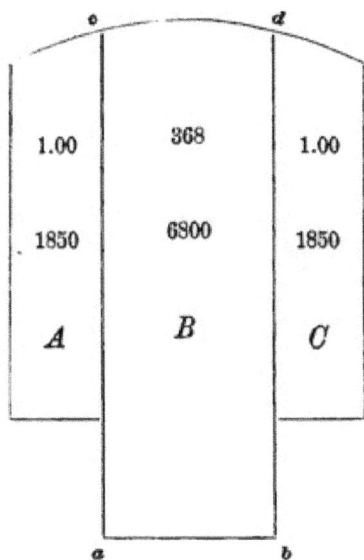

FIG. 1.

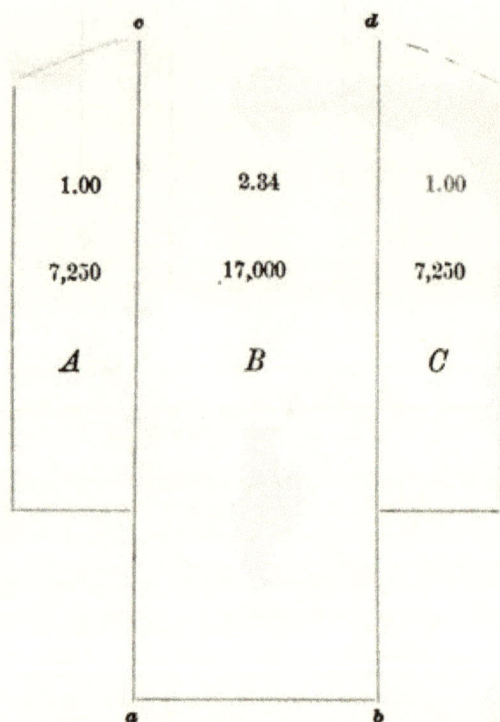

Fig. 2.

They are drawn to a common scale of four feet to an inch, and exhibit the relative size of a car at present in use on the ordinary gauges, (No. 2.) and a car such as is required on a 36-inch gauge. (No. 1.)

The smaller car is 7 feet outside width, with a clear hight of 7½ feet under centre of dome.

The small letters "a" and "b" represent the base on which the cars stand; that is, the distance from out to out of rail, which in the 3 ft. gauge is 40 inches, so that for this small car the overhang is 22 inches on a side.

Now, to compare the stability of cars on the two systems, let us divide each car by two vertical lines "a c" and "d b," projected upwards from the outside edge of the rail.

In the smaller car the outside sections A C weigh each 1,850 lbs., and the middle section weighs 6,800 lbs; that is, if A weighs 100 parts of the whole, B is 368 and C 100.

To secure this car against overturning, the 100 parts that are hung on either side of the rail are counterbalanced by the remaining 468 parts. Whereas, in the larger car, 100 parts of overhang are secured by only 334 parts of counterbalance; so that we find our small gauge car is really the safer and more stable of the two—chances of overturning are as 27½ to 42½, or thereabouts.

No wonder then, that the testimony of Captain

Taylor, Royal Inspector of Railroads in England, issued to the Festinoig railroad in Wales authority to run their passenger trains at any rate *they may desire* (although he limited them to 12 miles per hour when first opened, and still limits the standard gauge roads in that country) for he declares that he travelled over this little road at *thirty miles per hour* with every feeling of safety, and that a system of lines like this can be built costing but *two-thirds* of those now constructed and maintained at *one-half the expense* for any country.

## LOUISVILLE AND NASHVILLE RAILROAD.

### CONCURRING TESTIMONY.

Again, we have the testimony of J. P. Boyle, Esq., Chief Engineer of the Louisville and Nashville Railroad, in a report to his company in ilustration of the actual workings of wide and narrow gauges. He says:

The London & N. W. Railroad, England, has a gauge of 4 ft. 8½ inches. The business done amounts to about fifteen millions tons per year, five millions of which is mineral, and ten millions tons of general merchandise. With such a large movement of paying freight let us see what the movement of dead weight would be in handling this business:

As it has been demonstrated that the proportion of non-paying to paying weight transported over a wide gauge is at least 5 to 1, (and runs 7 to 1) the amount of dead weight hauled over the L. &. N. W. R. R., to transport the **ten** millions tons of freight would amount to fifty millions tons at an average speed of twenty-five **miles an** hour.

LET **US SEE** WHAT **A** WIDE GAUGE DOES WITH **IT.**

The entire length of the road named is 1,450 miles. The average gross weight of each train hauled is 250 tons, which requires 240,000 trains to haul the sixty millions tons required—or in 313 working days 767 trains per day over all parts of the road of 1,450 miles in 24 hours.

The company's books show that each net ton produces about $1.20, which, at 3 cts. per ton per mile, makes the average distance travelled by each ton of freight 40 miles, and consequently each train only averages 40 miles in distance per day travelled, (the Board of Trade—average being 23.⁴⁷₁₀₀.) The road being 1,450 miles long, it follows that there must be an average of 36 trains distributed along the length of the road constantly; this number divided into the total number of trains per day shows an average of about 21 trains per day passing over each mile of road, or one every 70 minutes. Thus, it will be seen, that notwithstanding the movement is so enormous, if the trains sustain an average speed of 25 miles per hour, **one** train following another is in

70 minutes behind the preceding one and in distance about 27 miles. It will thus be seen what a large surplus capacity this road has for doing business that is not utilized.

Let us now see what can be done with this business

## ON A 3 FEET GAUGE.

In the first place it is proven that a speed of 30 miles per hour can be maintained on a 3 ft. gauge, yet we will accept the same speed as on a standard gauge, that is, 25 miles per hour.

The narrow gauge freight cars weigh 1½ tons and will carry 4 tons of paying freight; ten million tons paying freight, therefore, would require three million seven hundred and fifty thousand tons dead weight to be moved, or a total of thirteen million seven hundred and fifty thousand tons gross weight moved on the narrow gauge to sixty million on the wide gauge.

We will assume that the narrow gauge trains each weigh 150 tons; it would require then ninety-one thousand six hundred and sixty-six trains each year to handle the amount of freight named, or in 313 days 284 trains per day; each train averaging 40 miles per day, the road being 1,450 miles in length, there would be an average of 36 trains distributable over the road daily in place of 21, as heretofore stated. This number divided into

the total number of trains per day, and we have
an average of eight trains every 24 hours passing
over every mile of the road, or one every three
hours, if the trains sustain an average speed of 25
miles per hour, one train follows another three
hours and seventy-five miles behind the preceding
one.

TABLE OF COMPARISON OF RESULTS.

|  | 4 FT. 8½ IN. GAUGE. TONS. | 3 FT. GAUGE. TONS. |
|---|---|---|
| Paying weight | 10,000,000 | 10,000,000 |
| Non-paying weight | 50,000,000 | 3,750,000 |
|  | 60,000,000 | 23,750,000 |
| Speed | 25 Miles. | 25 Miles. |
| Total wt. of each train | 250 Tons. | 150 Tons. |
| No. Trains per day | 767 | 284 |
| Length of road | 1450 Miles. | 1450 Miles. |
| One train every | 70 Minutes | 3 Hrs. |
| Distance apart in min'ts. | 70 " | 180 Min. |
| Do. Trains apart | 27 Miles. | 75 Miles. |

### SUPERIORITY OF THE NARROW GAUGE.

From the above it will be seen that the capacity
of the 3 ft. gauge for freight transportation over
the standard gauge is as 180 to 70, and that in
transporting the same amounts of dead weight as
the wide gauge, viz: fifty million tons, the narrow
gauge would move one hundred and thirty-six millions tons paying weight as against ten million

tons of paying weight moved on the standard gauge. These results are most astonishing, but if carefully examined will be found correct.

### HON. H. G. STEBBINS, VICE PRESIDENT SOUTHERN PACIFIC RAILROAD.

In the summing up of the able article written by this gentleman, he says:

"To sum up all, the narrow gauge system clearly has the advantage in these particulars. First—In the large comparative saving in first construction 'and right of way. Second—In the larger proportion of paying load to non-paying weight of train. Third—In the great reduction in wear and tear of permanent way, through the advantage gained by using lighter rolling stock. Fourth—In the great saving in the reduced wear and tear of wheels and tires, from the reduced weight on each wheel. Fifth—In the large proportionate increased power of locomotives (from the dead weight on the engines being used as tractive power. Sixth—In proportionate increased velocities gained by the light system. Seventh—In the greater economy in working traffic; and, Eighth—In the comparative increase in the capacities of traffic."

# DENVER AND RIO GRANDE RAILROAD.

## THREE FT. GAUGE.

The Denver and Rio Grande Railroad Company is practically demonstrating the superiority of the new system over the old, by having undertaken to construct a; 3 ft. gauge road over the almost entirely undeveloped country lying between the Union Pacific Railroad and the Rio Grande, a distance of 850 miles and upwards.

Eighty miles of this road is in successful operation from Denver to Colorado City, built at a cost not to exceed $13,500 per mile, and fully equipped.

The engines adopted by the company are of the same pattern as the standard engines on the Pennsylvania roads, with slight modifications.

The passenger engines have four drivers, 40 in. diameter, and one pair of leading wheels; cylinders 9 by 16 inches; weight on drivers in steam 25,000 pounds, with four-wheeled tender weighing 6,000 pounds empty.

Freight engines have six driving wheels coupled, and one pair leading wheels, cylinders 10 by 16 inches; weight on drivers 30,000 pounds; total weight, 35,000 pounds; tender weighs $6,000 pounds empty.

In the construction of the passenger cars there has evidently been too much deference paid to the prejudices in favor of the existing patterns, to obtain the full benefits of the narrow gauge system.

They are of the following dimensions:

Length, 35 ft.; over all, 40 ft.; width inside, 6½ ft.; outside, 7 ft.; hight above rail of floor beams, 2 ft. 3 in.; hight to top dome, 10 ft. 6 in.; hight centre gravity, 3 ft. 2 in.; angle of stability, 58 deg. 30 min.; weight, 13,000 pounds.

Some of these cars are handsomely furnished in the style of the Pullman car, and give the same seating room as the cars on the Pennsylvania road. They are thoroughly ventilated and comfortable, and eight wheeled, with an arrangement of springs which makes them easy on the track. They weigh 13,000 pounds—6½ tons.

Freight cars are made of the present pattern, are four wheeled and weigh about 4,000 pounds, with a carrying capacity of 10,000 pounds, or 5 tons.

A special corresbondent of the Omaha *Herald*, writing from Colorado Springs, December 24th, says:

## "NARROW GAUGE.

"The Denver and Rio Grande Railway (narrow gauge) remains open, notwithstanding the snow. It has not yet failed a single day in getting a train through between this place and Denver, although it crosses "the great Divide," next to the highest road pass in the world, and 200 feet higher than the Sierra Nevadas on the Central Pacific. The 3 ft. gauge is a complete success. The earnings of the first division of seventy-six miles of the road have averaged $3,000 weekly, about half for freight and half for passengers, in the seven weeks of almost continuous snow storms since its opening. The freight traffic offering, however, is double what the company can carry. They have ordered their rolling stock trebled immediately, and when this additional equipment arrives they will have eleven locomotives and 220 freight cars. The road will reach Pueblo, 120 miles from Denver, in March next, and Canon City by next summer."

COMPARISON BETWEEN THE PENNSYLVANIA AND DENVER RAILROADS.

The standard passenger engine on the Pennsylvania road weighs 40,000 lbs. (double that of the Denver road) and a total weight of 105,000 lbs., or 52½ tons. The usual load of passenger trains consists of

|  | TONS. |
|---|---|
| 4 Pass. cars seating 53 passengers each and weighing | 76 |
| 1 Baggage car | 14 |
| Engine and Tender | 52½ |
| Total | 142½ |

This train will accommodate 212 passengers, if full.

The same number of passengers on the Denver road require

|  | TONS. |
|---|---|
| 6 Passenger cars, weighing | 42½ |
| 1 Baggage car | 4 |
| Engine and Tender | 15½ |
| Total | 61 |

Or less than *forty-five-hundredths* the weight of the standard train to do the same business.

Sleeping cars, as comfortable in every respect as on the standard gauge, can be used on the road without difficulty.

In the freight train the comparison is about as follows:

| | WT. ON DRIVERS. | WT. OF ENGINE AND TENDER | | LOAD ON LEVEL EXCLUSIVE OF ENG'N & TEN'R. |
|---|---|---|---|---|
| | LBS. | LBS. | TONS. | TONS. |
| Freight Eng. standard gauge | 60,000 | 114,000 | 57½ | 1440 |
| Denver & Rio Grande, 3 ft. gauge | 30,000 | 41,000 | 20½ | 722 |

Of this load on full gauge road the proportionate weight of cars being 1 to 1 of freight, there is

|  | TONS. |
|---|---|
| Weight of cars | 720 |
| Of freight | 720 |

On the 3 ft gauge the proportionate weight of car to load is 1 to 2¼ of freight; we have

|  | TONS. |
|---|---|
| Weight of cars | 206 |
| Of freight | 618 |

Or nearly *nine-tenths* of the paying weight hauled by the large engine.

EXTRACTS FROM LETTERS

By J. D. Hoff to General C. B. Stuart, Consulting Engineer, New York, dated Denver Dec. 8th, '71.

Mr. Huff visited this road in company with company with other distinguished scientists from the Atlantic States during the late snow storms in December, and witnessed the progress of a train of fifteen cars on a small engine of $10\frac{1}{2}$ tons: He says:

"I found the heaviest grades 65 ft. per mile and the sharpest curves 9 degrees. The length to Colorado City is 26 miles; total cost, $13,500 per mile.

The Kansas Pacific railway, built by the same engineer, General Greenwood, cost $24,000 per mile, and the difference in grading is in favor of the latter road. The largest locomotive weighs $20\frac{1}{2}$ tons.

The passenger cars seat 34 passengers as comfortably as any cars in this territory, and, with two closets and stove, weigh only 13,000 lbs. Their freight cars weigh 3,500 lbs., and carry a load of 10,000 lbs. when full, or nearly *three pounds of load to one of car. One ton of coal runs a train of seventy-five tons of freight eighty miles in four hours.* Their books show that the round trip of 152 miles costs about *fifty dollars ;* or $2\frac{1}{3}$ *tons one mile for one cent.* The trains went through the deep snow, beating all the other roads this winter.

The divide over which the train runs is 7,000 feet above tide water, and while the U. P. and the K. P. roads in the same territory have been blockaded for weeks at a time, this little road had not lost a trip up to the middle of December, and was never more than two hours behind time on any trip during the winter.

## ORANGEVILLE RAILROAD.

The Orangeville *Advertiser*, Canada, in speaking of the Toronto & Bruce narrow gauge says:

"The amount of goods which is brought is truly surprising, and the number of passengers travelling both ways is also very large. The fact is, the railroad is a great success; having gone to Toronto and back on the line the bresent week, we were agreeably surprised at the comfort of the journey. We have heard a good deal of the "narrow gauge," the " wheelbarrow railway," but let anybody get into the cars without being told anything about narrow gauges, and we venture to say that he wonld not observe the difference between it and any other railway. The cars are seated in the same way as the wide gauge, each seat accommodating two comfortably ; the track, too, is very smooth.

## RUSSIAN SYSTEM.

The system of roads in this country have been reduced from 6 ft. and under to the present standard of 3 ft. 6 in., carrying regularly 354 tons of train, exclusive of engine and tender, on grades of 1 in 85, some of which are five miles long. Total freight carried per annum, 376,430 tons; total number of passengers carried per annum, 189,762, at a cost of 1½ cent per mile, and freight at 2 cts. per mile.

## HORATIO SEYMOUR ON NARROW GAUGE.

First—That a 3 ft. gauge can be built for 60 per cent. of the cost of the standard gauge, and cost but one-half to run it.

Second—That the Festinoig railroad in 1868 carried 130,000 tons of freight and 145,000 passengers, and only 13 miles in length, while the Syracuse and Binghamton road, 81 miles long, only carried 424,537 tons and 245,577 passengers, and the Black River and Utica road carried 100,-111 passengers and 25,403 tons of freight, and 86 miles in length; while the Ullinbord road in Sweden, 23 miles long and 3 ft. 7 in. gauge, carried 100,000 passengers and 150,000 tons freight with 12-ton locomotives and a speed of 35 miles per hour.

Thirdly—That the narrow gauge system will not cost over two-fifths the present, and fully meet all the requirements of the largest business of any road in the country.

## CONCLUSIONS OF MR. SPOONER ON FESTINOIG AND THREE FT. GAUGE.

First—That through the whole time he has had the control of the Festinoig railroad it has entirely demonstrated the theory of the immense saving on the narrow gauge system: in having carried more freight and passengers at less cost than any line of railway now in use; that it is almost free from oscillation; that it has withstood the severest wind storms in the country without being affected; that the cars can run at 35 miles per hour with perfect safety; that the wear and tear of rolling stock and rails is reduced to an absolute minimum.

Secondly—That a 2 ft. 9 in. to 3 ft. gauge meets the only objection that can now be raised against a narrow gauge, and that all the requirements of commerce can be fully transacted by lines built on that gauge; that they can be built for from one-fourth to two-fifths of the cost of the standard gauge through the same section of country, and can be maintained at not to exceed one-half of the cost of the present system to do the same business.

## MR. FAIRLIE ON A SYSTEM OF RAILROADS FOR INDIA IN 1871.

The gauge established by law for that portion of the Dominion is 5 ft. 6 in. in width, which is now changed to 3 ft. and 2 ft. 9 in.

The arguments that contributed largely to the change were advanced by Mr. Fairly, in which he proved his case by taking the London and North-western Railroad, in England, the business of which being the largest in that country, and because its management is such that any shortcomings in its operation is wholly due to its construction and not to its management, and proceeds to show that if its gauge had been 3 ft. instead of 4 ft. 8½, its gross traffic could have been done at one-half its present cost, with half the motive power, and in such a way as to reduce the tonnage over the road one-half, and remove the necessity of the heavy expense that is incurred in the construction of a third line of rail. He says:

" The goods and mineral trrffic on the L. & N. W. road for a single year amounted to 15 million tons. I will assume that from out of the 15 million tons 5 are minerals consisting chiefly of coal, and deal only with goods which are left as the net revenue of the year's business.

"It has been proven that the proportion of non-paying to paying load is about 7 to 1; this would give 70 million tons of rolling weight employed in carrying ten million tons of paying load, or, in order to avoid all risk of exaggeration, I will assume the dead weight to be only as four to one; this reduces from seventy to forty million weight of wagons employed to carry the ten million tons of paying load. The whole gross load will then be fifty million tons hauled at an average speed of 25 miles per hour.

"The earnings of the goods traffic on this line sre 6s 3d (currency $1.50) per train mile run, which at an average rate all round of 12d per ton per mile would give about 50 tons as paying weight and 255 tons as gross weight hauled per train mile (Russian road 345.) Dividing this 255 tons into the 50 millions gives 196,078 trains; dividing this by 313 working days in a year, gives 626 merchandise trains on all parts of the road in 24 hours. This makes the average distance traveled by each ton to be about 38 miles; so that as each ton of the total weight hauled runs 38 miles, and the entire length of the road is 1450 miles, it follows that there must be on an average 37 merchandise trains distributed over the total length. This number divided into the total number of trains per day of 24 hours gives an average of over 17 trains per day passing over each mile of road. My object in bringing the figures to this point is to show that although at first sight the number of 626 trains per day looks large yet when divided over the entire line is comparatively small.

Having arrived at this conclusion, we are in a position to see how it would affect the question if the gauge was 3 ft. instead of 4 ft. 8½ inches.

In the first place, the same or a greater speed can be obtained, say up to 35 or 40 miles per hour. The speeds in each case therefore being equal, the next point to examine is the result of the carrying on the narrow gauge. The proportion of non-paying to paying load has been reduced from 7 to 1, to 4 to 1, although the former is the actual proportion. The wagons employed average four tons in weight, so that each wagon carries one ton for every mile it runs.

The wagons on a 3 ft. gauge weigh each a ton and carry a maximum load of three tons. Supposing that the same number run on the narrow as on the broad gauge it follows that the average ton of merchandise now carried would easily be carried in a wagon weighing one ton instead of four tons, and that the gross load passing over the line for one year would be only twenty millions of tons instead of fifty, while the same amount of paying weight would be carried in either case; that is, the small wagons, which are capable of carrying three times the weight of goods now actually carried on a four ton wagon would only have to carry one-third of that quantity, and produce the same paying load as the heavier wagons; thus, instead of fifty million tons travelling over the line there would only be twenty millions, and as the handling costs the same, whether paying or non-paying, it follows that the expense would be reduced *two-fifths* of what it now is.

We must also consider the enormous saving in permanent way which would have to bear the friction and weight of twenty millions tons instead of fifty millions. If we assume the same number of trains per day, the weight of each would be reduced from 255 tons to 102 tons; or if the same gross weight was employed, the number of trains per day would be reduced from 626 to 250. If there should be sufficient traffic to load the narrow gauge cars in such a way as to require the same number and weight of trains that are now worked, the result would be that—without increasing the cost of hauling or permanent way expenses—the 3 ft. gauge would carry a load of twenty-five millions tons as against the ten millions now carried.

Now, then, we have established the fact that so far as capacity goes, the narrow gauge is superior to the wide gauge.

The former can produce 25 millions net out of 50 millions tons, while the latter, to produce the same result, if continued to be worked as it is now, would require that 125 millions tons should be hauled, at an increased cost in the same proportion of 125 millions to 50 millions.

The gauge now established by law in India is 2 ft. 9 in., to be so arranged as to use the 5 ft. 6 in. machinery until new narrow gauge machinery may be required."

# THE ULLENBORG ROAD IN SWEDEN.

## GAUGE 3 FT. 7 IN.—LENGTH 37 MILES.

This road reports a business of 220,000 passengers and 327,000 tons of freight per annum, carried at per passenger $2\frac{1}{16}$ cents per mile, and for freight at $2\frac{5}{16}$ cents per mile per ton, at speed of 35 miles per hour, with a 12 ton locomotive, revenue about 15 per ct. per annum.

Major Adelskold, State Engineer of Sweden, says :

" In every case where small gauge railroads have been built they have realized every expectation, and I deem it a waste of money to build broader gauge when a narrow and cheaper gauge will meet all the requirements of the business of the country.

## SPECIAL CONCLUSIONS FROM THE ROADS ALREADY BUILT.

### FESTINOIG RAILWAY—$13\frac{1}{2}$ MILES, $23\frac{1}{2}$ IN GAUGE.

The results of operations of this road are

Passengers transported yearly, 140,000 passengers at 1 ct. per mile.

Freight hauled, 500,000 tons at $\frac{1}{2}$ ct. per ton.

Original cost, $180,000.

Present value, $500,000.

Interest on original investment, 40 per ct. per annum.

Interest on present value, 23 per ct. per annum.
Maximum gradients, 1 in 67.

"      Curves, 132 ft. radius.

"      Weight of Engines, 10 to 19½ tons.

Proportion of non-paying passenger traffic, ¼ ton per passenger.

Proportion of paying to non-paying freight, 3 to 1 per ton.

Freight cars 1,344 lbs., carrying 3½ tons.

## . THE LAREN ROAD IN NORWAY.

GAUGE 3 FT. 7 IN.—LENGTH 73 MILES.

Cconstructed by Mr. Pihl at a cost of $25,000 per mile. Reports annual income of 15 per cent. on the cost, at 3 cents per passenger per mile, and 2¾ cents per ton per mile. Locomotives, 18 tons, grades, 1 in 62 ; curves, 237 ft. radius, speed, 35 miles per hour.

The Hamer road in Norway, 30 in. gauge, constructed by Mr. Pihl at a cost of $15,000 per mile. Rails 30 lbs. per yard, locomotives 12 to 20 tons, grades 1 in 80, and curves 400 ft. radius, at 3 cents per mile per passenger, and $2\frac{5}{16}$ cents freight per mile carried ; annual dividends, nearly 17 per ct. Speed 26 miles per hour.

Cost of standard gauge in the same country, $32,000 per mile, pays at same rate of charges less than 8 per cent. per annum.

# SOUTHERN PACIFIC RAILROAD.

## WHAT CAN BE DONE THERE.

We will now take the statistics of the S. F. &
S. J. R. R. Co. We will suppose that this company runs two passenger trains per day each way
with four passenger cars each, the average weight
of each train will be as follows:

|  | TONS. |
|---|---|
| Locomotive and Tender | 32 |
| 6 Passenger Cars, light | 60 |
| 27 Passengers to each 162 Pass | 12¼ |
| Baggage Car and Load | 11½ |
| Express Car and Load | 12¼ |
| Total | 128½ |

Average weight on a 3 ft. gauge to do the same
business:

|  | TONS. |
|---|---|
| Locomotive and Tender | 10 |
| 9 Passenger Cars carrying 18 Passengers each and weighing 4 tons each | 36 |
| 162 Passengers | 12¼ |
| Baggage Car and Load | 2¼ |
| Express Car and Load | 2¼ |
| Total | 63½ |

Or less than one-half the train load that is now
carried.

It will be seen that of these two trains the
passenger weight of 12½ tons, and their baggage

1½, and the express matter of about half a ton, amounting to 14½ in all, is the paying freight of the whole train, which weighs 128½ tons. The proportion, therefore, is nearly 8 tons of non-paying weight to one ton of paying weight.

While in the 3 ft. gauge train we have the 14½ tons of paying weight, as before, to 63½ tons total train weight hauled, or less than 1¾ tons of dead weight to one ton of paying weight.

Apply the same system of construction to the freight business, and let us see what is the result:

The capacity of a ten ton engine on the grades

of this road would be......................295 tons
Deduct engine and tender.................... 19 "
_____
Leaves for load...........................276 tons

Cars carry 10 tons and weigh 2 tons, and to carry 276 tons would require 27 cars at 2 tons each; leaving for paying load, 222 tons.

We have 222 tons of paying load to 54 tons of cars, or over 4 tons of paying load to one ton dead weight.

SAN FRANCISCO AND SAN JOSE RAILROAD FREIGHT.

We will suppose the standard gauge freight engines on the San Francisco and San Jose Railroad has double the weight on the drivers and double the hauling capacity, or 590 tons; deduct engine and tender, 40 tons; leaving load, 550 tons.

The freight cars weigh from 8 to 10 tons, and carry 10 tons. One half will then be cars, 275 tons ; leaving freight, 275 tons.

This shows that the small engine and cars of the 3 ft. gauge has a hauling capacity of 224 tons for a 19 ton engine, against 275 tons on the present road for a 40 ton engine, or over two to one of hauling capacity.

Follow this to its financial conclusions, we have the following result :

Passenger traffic on one train per day each way, 324 passengers carried 50 miles per day, 16,200 miles ; or per year, carried one mile, 5,913,000 miles. Total cost transporting passengers at $2\frac{40}{100}$ cents per mile, $141,912.

## ON THE THREE-FOOT GAUGE,

With he proportions of 8-1½, we have of the above business, total miles run as above, 5,913,000 ; total cost, $31,043 ; leaving a saving of $110,869—or about one-fifth of the present rates of charge, which would be 50 cents from here to San Jose.

The saving to the farm produce and of freight would equal or exceed the passenger traffic in the same proportion, as the passenger machinery and rolling stock recedes in weight from that of freight machinery and rolling stock.

## GENERAL RESULTS.

**It will be** admitted that **the chief** difficulty in building railroads anywhere **is** their cost. **This** difficulty hinders **the** building of many roads **that** would pay if they were built.

The present standard gauge railroads in the **United** States **have cost** about $44,000 per mile, the construction of which has created an indebtedness **of** nearly 2,200 **millions** dollars (exclusive of stock subscription) which at 7 per cent. interest per annum involves **the annual payment of** 154 millions dollars which the commerce of **the** country **is** compelled **to** pay. If the present system is **not changed, and the** bond list increases in the **next ten years** as it has **in the** last, this indebtedness **will amount to probably** not less than **3,500 millions** dollars, with **an** annual **tax on transportation** of 245 millions dollars.

The change that is proposed will build roads that have an equal capacity, both for freight and passenger, and upon which equally fast time can be made, for less than 40 per cent.; and when built can be maintained **at not to** exceed one-third of the expenses of the present system. Thus two of the most formidable objections in the way of construction of railroads in sparsely settled countries would be removed—that of an exorbitant cost and **an** expensive system **of** management. There is scarcely a county in this State that could not afford to build a small gauge railroad wherever a good public highway was necessary. Even the lines already occupying our principal avenues of business would find it necessary to change the present standard or be compelled to reduce their charges to such low rates as not to be able to pay running expenses.

The cutting and filling of the road bed, will not be over one-half through a country of an ordinary even surface, curves can be used and successfully operated as sharp as 200 ft. radius, and thereby reduce the cost of construction in a heavy country to less than *twenty-five per cent. of what a standard gauge can be built for in those countries.*

The cost of permanent structure can be reduced at least one-third; ties to two-thirds. The rails will be reduced from 80 lbs. per yard, to 25 and 30, and their durability increased, from five years to at least fifteen, with the same business.

The locomotives can be reduced in weight from 20 and 60 tons, to 6 and 15 tons, made on patterns suitable to the changed gauge.

The cars will be reduced from 16 and 30 tons to 4 and 6½ tons, while the freight cars can be reduced from 6 and 14 tons, to 1 and 2½ tons.

The buildings can be reduced in proportion. Turn-tables and all fixtures can be reduced at least three-fifths.

This system of roads can be constructed in one-half the time required to construct the present system. and return interest proportionately enhanced.

The principal cost of maintaining a road depends upon the weight the road bed has to carry to do its business, the proportion being as three to one of all other expenses. This is reduced in passenger traffic from 1 ton and $3\frac{1}{2}$ tons dead weight for each passenger; and of freight, from 7 tons dead weight to one ton paying weight, on the standard gauge; to one-quarter of a ton dead weight for passengers; and to one-third ton of dead weight to 1 ton to paying weight for the proposed narrow gauge system.

The present system as a result, has given over 60,000 miles of railroads, on which we are obliged to carry at least 6 tons of carriages to one ton of freight, and from 10 to 30 tons of carriage for one ton of passengers, costing for its transportation two hundred millions dollars annually on bonds more than is necessary, and getting in return the satisfaction of knowing that we are expending millions yearly in manufacturing rolling stock for the special purpose *only* of wearing out the rails it runs on; while the practical result of the proposed change will be, economy in coustruction and operation, cheap fretghts, large dividends, adaptability to other than mountainous countries, feasibility of

doubling the number of miles of our railway systems at one-third of the former outly, the easy development of sparsely settled but valuable districts
and the long train of advantages, which is sure to
follow to the stockholders who invest in those
roads, as well as to the people who will by their
agency be induced to occupy our unsettled regions.
And above all, and before all, other considerations,
they will constitute a means through which existing lines of standard gauge roads will be compelled
to respect the convenience and necessities of the
public, in reduction of freights and charges, from
six cents a mile per passenger to less than two,
and from ten cents per ton for freight, to at least
one-third that amount.

Extract from the Reports of the Joint Committee of the Legislature of the State of Massachusetts, March 11, 1871.

Estimated cost of building one mile of narrow gauge railroad, where the fills and cuts are 4 feet:

| | |
|---|---:|
| Rails.................................................$ | 4,243 |
| Ties,............................................... | 352 |
| Spikes.............................................. | 175 |
| Fish Joints, Bars and Bolts.................... | 400 |
| Laying Track...................................... | 250 |
| Embankments, 6,062 yards.................... | 1,513 |
| Cuts, 5,629 yards................................ | 1,480 |
| Rock Cut, 1,611 yards............,............ | 1,611 |
| Ballast, 1,000 yards............................. | 1,000 |
| Sidings ............................................ | 200 |
| Masonry and Bridges........................... | 1,140 |
| Total........................................$ | 12,364 |

Estimate of same road, with similar grades and alignment, of a 4 ft. 8½ in. gauge.

| | |
|---|---:|
| Rails.................................................$ | 6,600 |
| Ties ................................................. | 924 |
| Spikes............................................... | 264 |
| Fish Joints, Bars and Bolts.................... | 700 |
| Laying Track...................................... | 325 |
| Embankment, 8,604 yards.................... | 2,151 |
| Excavations, 11,703 yards.................... | 1,927 |
| Rock Cuts, 2,085 yards........................ | 2,085 |
| Ballast ............................................. | 2,000 |
| Sidings.............................................. | 334 |
| Masonry and Bridges........................... | 2,000 |
| Total........................................$ | 19,301 |

Table of the Principal Dimensions and Weights of Various Patterns and Sizes of Narrow Gauge Locomotives. With Loads they will have on a Straight Track in good condition.

| KINDS OF LOCOMOTIVES. | CYLINDER. diam. in. | CYLINDER. stroke in. | DIAMETER OF DRIVERS, in. | WEIGHT IN WORKING ORDER. TOTAL LBS. | ON DRIVERS. | ON EACH PAIR OF DRIVERS. | LOAD IN GROSS TONS IN CARS. LEVEL, tons. | 40 FT. GRADE | 80 FT. GRADE | 100 FT. GRADE |
|---|---|---|---|---|---|---|---|---|---|---|
| **CLASS NO. 1.** | | | | | | | | | | |
| Four Wheeled Connected | 9 | 12 | 30 | 18000 | 18000 | 9000 | 390 | 120 | 70 | 55 |
| Tank Locomotive | 9 | 16 | 26 | 22000 | 22000 | 11000 | 480 | 150 | 85 | 70 |
|  | 10 | 16 | 36 to 40 | 36000 | 26000 | 13000 | 590 | 180 | 105 | 85 |
| **CLASS NO. 2** | | | | | | | | | | |
| Four Wheeled Connected | 9 | 12 | 30 | 16000 | 16000 | 8000 | 385 | 115 | 95 | 50 |
| Do with separate Tender | 9 | 16 | 36 | 20000 | 20000 | 10000 | 480 | 140 | 75 | 60 |
|  | 10 | 16 | 36 to 40 | 24000 | 24000 | 12000 | 380 | 170 | 95 | 75 |
| **CLASS NO. 3.** | | | | | | | | | | |
| Six Wheeled Locomotives | 10 | 16 | 36 | 28000 | 28000 | 9333 | 615 | 185 | 107 | 85 |
| Do Tank Engine | 11 | 16 | 36 | 33000 | 33000 | 11000 | 740 | 225 | 130 | 105 |
|  | 12 | 16 | 36 to 40 | 38000 | 38000 | 12666 | 860 | 265 | 150 | 125 |
| **CLASS NO. 4.** | | | | | | | | | | |
| Six Wheeled Locomotives | 10 | 16 | 36 | 25000 | 25000 | 8333 | 605 | 175 | 95 | 75 |
| Do with separate Tender | 11 | 16 | 36 | 30000 | 30000 | 10000 | 730 | 215 | 120 | 95 |
|  | 12 | 16 | 36 to 40 | 35000 | 35000 | 11666 | 850 | 255 | 140 | 115 |
| **CLASS NO. 5.** | | | | | | | | | | |
| Four Wheeled Locomotives | 9 | 16 | 36 to 40 | 25000 | 20000 | 10000 | 480 | 140 | 75 | 60 |
| 1 Pair Leading Wheels | 10 | 16 | 36 to 40 | 30000 | 25000 | 12500 | 605 | 175 | 95 | 75 |
| **CLASS NO. 6.** | | | | | | | | | | |
| Six Wheeled Locomotives | 10 | 16 | 36 | 30000 | 25000 | 8333 | 605 | 175 | 95 | 75 |
| 1 Pair Leading Wheels | 11 | 16 | 36 to 40 | 35000 | 30000 | 10000 | 730 | 220 | 120 | 100 |

# COMPARATIVE WEIGHTS OF NARROW AND FULL GAUGE CARS.

| | WT. OF CAR EMPTY. LBS. | CAPACITY. LBS. | WT. OF CAR LOADED. LBS. | WT. OF WHEELS AND AXLES. LBS. | WT. OF ALL THE WHEELS. LBS. | WT. OF EACH WHEEL. LBS. | PER CT. DEAD WT. TO LOAD. |
|---|---|---|---|---|---|---|---|
| Full Gauge 8-Wheel Box Car.... | 20,000 | 20,000 | 40,000 | 5,400 | 34,600 | 4,325 | 100 |
| Narrow Gauge 4-Wheel Box Car.. | 4,500 | 8,000 | 12,500 | 1,200 | 11,300 | 2,825 | 56 1-2 |
| Full Gauge 8-Wheel Flat........ | 16,000 | 24,000 | 40,000 | 5,400 | 34,600 | 4,325 | 66 2-3 |
| Narrow Gauge 4-Wheel Flat...... | 3,500 | 9,000 | 12,500 | 1,200 | 11,300 | 2,825 | 38 8-9 |
| Full Gauge 4-Wheel Coal Car.... | 6,600 | 11,000 | 17,600 | 2,300 | 15,300 | 3,825 | 60 |
| Narrow Gauge, 4-Wheel Freight.. | 4,000 | 8,500 | 12,500 | 1,200 | 11,300 | 2,825 | 47 |

www.ingramcontent.com/pod-product-compliance
Lightning Source LLC
Chambersburg PA
CBHW021952190326
41519CB00009B/1232